Guidelines for selection of
biological SSSIs:

bogs

Copyright Joint Nature Conservation Committee 1994
ISBN 1 873701 70 5 - looseleaf
ISBN 1 873701 71 3 - bound

Date of publication: December 1994
Printed by Stephen Austin and Sons Limited

Foreword

Guidelines for selection of biological SSSIs: bogs

These *Guidelines* supersede those given in Chapter 8 of *Guidelines for selection of biological SSSIs* (Nature Conservancy Council 1989).

This document should be used in combination with the original *Guidelines for selection of biological SSSIs* (Nature Conservancy Council 1989) in which there is an explanation of the rationale behind the SSSI system and of terms such as Areas of Search.

These guidelines only apply to sites notified as SSSIs following publication of these guidelines. They are not intended to apply where the notification procedure for a site has already been started at the time of their publication.

Reference

Nature Conservancy Council. 1989. *Guidelines for selection of biological SSSIs.* Peterborough.

This document should be referenced as:

Joint Nature Conservation Committee. 1994. *Guidelines for selection of biological SSSIs: bogs.* Peterborough.

Bogs

1. Introduction

1.1 Ombrotrophic (rain-fed) mire, so called because its mineral nutrients are derived principally from rainfall rather than ground-water sources, is the other main class of peatland. In Britain these mires are termed **bogs** and in contrast to **fens**, which are fed by mineral-enriched waters, their vegetation is characterised by acidophilous plant communities in which the genus *Sphagnum* usually is, or has been, a conspicuous component. In the humid, oceanic climate of Britain, ombrogenous (rain-generated) bogs are an especially important element of the original range of vegetation formations. They are an extensive feature of western and northern areas, where measurable rain falls on two days out of three. This regular input of precipitation produces a fairly constant level of surface seepage on many bogs, which gives rise to other distinctive communities which in Fennoscandia would be regarded as fen (minerotrophic mire) but in Britain are considered to be part of bog complexes. In lowland areas with predominantly acidic substrata there are examples of valley and basin mires which receive acidic surface seepage giving rise to ombrotrophic vegetation similar to that of ombrogenous mire. These may be classified as fen/bog complexes (see Chapter 7, section 4, NCC 1989).

1.2 Topography as well as precipitation plays a key part in the development of ombrogenous bogs. These peatlands are limited to flat or gently sloping terrain, which promotes the essential degree of substrate waterlogging by holding rainwater. As gradients of climate increasingly create a water surplus by causing the ratio of precipitation to evaporation to rise in a north-westerly direction in Britain, so the conditions for bog formation become more favourable. A prevalence of hard, acidic rocks and base-deficient soils also favours the development of acidophilous bog vegetation, while the cool humid climate promotes leaching and podsolisation, which reinforce these edaphic tendencies. The peat formed under these conditions is nutrient-poor and low in pH, and may become even more acidic when drained. The soils that form pose considerable constraints on agricultural utilisation and are generally unsuitable for arable cultivation, in contrast to fen peat soils.

1.3 There are two main types of ombrotrophic mire, **raised bog** and **blanket bog**. **Raised bog** is characteristic of relatively flat underlying topography or basins, and so found mainly on low plains or broad valley floors (Figure 1). They may overlie sites of shallow glacial lakes which became infilled and occupied by fen. In subsequent developmental stages expanding nuclei of acidophilous vegetation spread gradually over the whole surface. Raised bog types may therefore contain evidence of earlier fen phases of development (see Chapter 7, NCC 1989) within their stratigraphic record. Raised bogs form under a range of conditions from strongly oceanic to relatively dry continental and they are widespread in Boreal regions.

Unlike blanket bogs, the surface contours of raised bogs are generally independent of the underlying mineral topography. The typical raised bog has a gently domed profile, with peat depth greatest in the centre and then decreasing gradually, with the edges marked by a steeper rand (sloping mire margin). This is sometimes bounded by a minerotrophic fen or lagg, often drained by a small stream. This structure is seldom seen in British raised mires because the majority have been widely modified hydrologically and physically by draining, fire, afforestation and peat-cutting. Many have been totally removed for agriculture, and even the best remaining examples have often undergone considerable modification of their margins by peat-cutting and agricultural reclamation.

Intact bogs have up to 98% of their structure in the form of water, with only 2-5% by weight as solid material. Water levels within the dome are held above those of the surrounding, regional water table. In some cases this perched water table may attain heights of 9-10 m above the regional water table.

1.4 **Blanket bog** is a vegetation formation which probably reaches its extreme world development in western and northern Britain and in western Ireland, reflecting the cool, intensely oceanic climate (Figure 2). The conditions necessary for soil waterlogging favourable to *Sphagnum* growth and peat formation are such that bog development is no longer confined to level terrain, but can occur on all but the more steeply sloping ground. It therefore covers many of the gentler uplands in a smothering mantle; hence the descriptive name.

In the north-west Highlands significant peat development can be found on slopes of up to 30 degrees on shady aspects, and in the wettest parts of Scotland and Ireland chemical/edaphic limitations can be so overridden that acidic peat forms even over limestone.

Unlike raised bog, blanket bog generally mimics the more variable underlying topography upon which it forms. Peatland vegetation in blanket bogs often grades imperceptibly into that of drier mineral soils through transitional wet heath or grassland types and less typically ends abruptly with a distinct margin. Peat depth in blanket bogs is usually related to topography.

1.5 Discrete areas of raised bog and blanket bog may both occur in some districts. In many areas peatlands which may have begun as raised bog have become swallowed up in a general expanse of blanket bog, losing their distinctive marginal features. In such circumstances it is generally not practicable to separate these elements from the general blanket bog classification. In some regions, too, there are areas of bog types intermediate between raised and blanket bog, usually at elevations between 30 m and 150 m (see section 5).

1.6 Ombrotrophic mires have great value to Holocene ecological and archaeological studies (e.g. Godwin 1975). Raised bogs tend to have longer, and therefore more important, stratigraphic sequences than blanket bogs, because of their previous fen and lake phases, and in some sites these sequences go back to the early part of the late glacial period or possibly even interglacial periods. Blanket bog formation is generally believed to date from the wetter climatic conditions at the onset of the Atlantic period about 7,500 years ago, but in many situations peat formation did not begin until much later (Birks 1988). Blanket bog peats thus generally contain a shorter post-glacial record of vegetational, climatic and other history than raised bogs, but their more extensive occurrence gives the opportunity of studying a wide range of local conditions (e.g. Bridge *et al.* 1990; Birks 1975; Gear & Huntly 1991). While the extent is imperfectly understood in Britain, blanket peat can conceal important prehistoric landscapes and field systems (such as the Ceide fields, County Mayo, Eire).

1.7   Bog vegetation is characterised by a dominance of acidophilous plants. In the wetter periods of climatic cycles, undisturbed surfaces may be an almost continuous mixed carpet of *Sphagnum* species. Surfaces in drier climatic periods and eastern locations, or those which have been disturbed, may have less *Sphagnum* and a greater abundance of vascular plants such as *Calluna vulgaris, Erica tetralix, Eriophorum vaginatum, E. angustifolium* and *Scirpus cespitosus* rooted in the upper peat or living moss. In the north and west of Britain, the most natural, undisturbed bog surfaces usually display a distinctive microtopography ('hummockiness'), showing a fine-scale vertical zonation of vegetation between wetter hollows and drier hummocks. Some bogs show even greater surface differentiation into deep open-water pools and more pronounced, and hence drier, hummocks. There is a great deal of variation in the nature of the hummock-hollow patterns encountered across Britain. Patterned bog surfaces are especially well developed in Scotland and are an intriguing aspect of the long-term upward growth and peat formation in these ecosystems (sections 3.1.4 and 3.1.5). The distribution of National Vegetation Classification (NVC) types within the microtopographical zones is presented in Table 1. Because the National Vegetation Classification (Rodwell 1991) does not deal fully with the fine-scale pattern of plant communities on ombrogenous bogs, the classification has been refined as shown in Figure 3.

1.8   Intact ombrotrophic bogs are dependent upon climate and, particularly in the case of blanket bogs, topography for their overall structure, surface pattern and even, to a large degree, species composition. Consequently, if the full range of variation within and between bog systems is to be conserved, it is essential that attention is paid to local, regional and national climatic patterns. As these factors vary greatly across Britain, it is important to encompass a wide geographic spread in site selection. "Mires are a zonal phenomenon. . . . they are affected by local differences in natural conditions, among which climate and relief play a leading part." (Ivanov 1981, p. 7).

1.9   Peatlands have been much used for agriculture and forestry, and as a source of fuel. Many have been drained and exploited commercially for horticultural peat and moss litter in recent decades. Bogs have a low carrying capacity for large vertebrates, but the sheer extent of blanket bogs in particular makes them an important part of the grazing range of many sheep walks, deer forests and grouse moors. The widespread phenomenon of 'hagging' of many blanket bogs (erosion by gullying) is often attributed to the combined effect of grazing and burning. Fire is often implicated for the widespread drying of the surface of bogs, attributed to a reduction in *Sphagnum* cover. The status of some erosion features is still far from clear, and there may be examples which represent a natural phase in bog development (Stevenson *et al.* 1990).

1.10　　A high proportion of Britain's original bogs have been altered in these ways, making ombrotrophic bogs among our most threatened natural habitats. The estimate in the National Peatland Resource Inventory being prepared by Scottish Natural Heritage is that, of the 69,000 ha of raised bog soils in Britain, only some 3,826 ha (6%) remain in a near-natural state. Table 2 shows the distribution of the remaining lowland raised bog resource in Britain. The inclusion of these habitats in Annex 1 of the Habitats and Species Directive (92/43/EEC) is a recognition of the rarity and fragility of ombrotrophic bogs within the area of the European Union (see Annex 1). Britain contains a very high proportion of the remaining bogs in the European Union, 'active' bogs being afforded priority habitat status under the Habitats and Species Directive (HSD). The agreed definition of 'active' for the purposes of the HSD states that the site should be "still supporting a significant area of vegetation that is normally peat forming, but bogs where active formation is temporarily at a standstill, such as after a fire or during a natural climatic cycle, e.g., a period of drought, are included".

1.11　　In biodiversity terms, bogs are especially important as repositories of highly specialised species and assemblages well-adapted to waterlogged and nutrient-poor environments.

When selecting sites, particularly raised bogs, for SSSI notification it is important to take fully into account the presence of rare species and vegetation types given specific statutory protection.

1.12　　Lowland raised bogs are by nature separated from each other, often by wide tracts of land inhospitable to bog species. This natural fragmentation has been exacerbated by the large reduction in numbers of near-natural sites witnessed over the last half century.

Fragmentation is known to put at risk the survival of the full species complement of the biotope. The full biodiversity of the raised bog resource surviving in Britain may therefore be regarded as fragile. This may be particularly so for the relatively immobile flora and invertebrate fauna. For these reasons alone the area of extant natural lowland raised bog may be close to or even below the 'critical natural capital' required to guarantee the survival of all species and species assemblages characteristic of this biotope in Britain.

2.　International importance

2.1　　Raised bogs are a relatively widespread peatland type in the northern hemisphere, but the British examples are important structurally and floristically as extreme oceanic examples with features peculiar to this country. Bogs of similar type were once widespread in Ireland, but many have been completely destroyed, and most of the remainder significantly altered.

2.2   Britain is one of the main world locations for blanket bog and contains about 10-14% of the total global area of this extremely restricted vegetation formation (Lindsay *et al.* 1988).

2.3   Blanket bog is a tundra type of ecosystem occurring in Britain in a southern and insular context, forming as a response to cool oceanic conditions, as distinct from the permafrost environment which creates true tundra.

2.4   The floristic composition of blanket bog and associated wet heath in Britain is unique in the world and demonstrates a highly Atlantic influence on plant distribution and vegetational development. Even the similar Irish examples differ in important floristic features.

2.5   Blanket bogs in northern Scotland are notable for the occurrence of a tundra-type breeding bird assemblage showing general similarity to, but specific differences from, that occurring on Arctic and Sub-arctic tundra. British blanket bogs support significant fractions of the total populations of certain breeding bird species in Europe and particularly in the territories of the European Union (Stroud *et al.* 1990). (See also Chapter 14, NCC 1989).

2.6   Because the surface features and vegetation of bogs throughout their history have been intimately linked with the prevailing climate, the bog archive conserved in peat layers has become increasingly used in international studies of climate change (Barber 1981). The conservation of this resource is particularly important in the light of current concern about global warming.

3.   Selection requirements

3.1   In order to determine the range of bog variants within a particular Area of Search (AOS), it is first necessary to consider the variations in structure of ombrotrophic mires. Five structural levels are recognised (Lindsay *et al.* 1988; Lindsay in press), on a hierarchical scale of decreasing size, as follows (see Figures 2, 4 and 5).

         3.1.1   Mire **macrotope**, large-scale units, consisting of complexes in which peat bodies originating as different hydrological units have become either closely juxtaposed or merged together, e.g. the Silver Flowe in Galloway.

         3.1.2   Mire **mesotope**, in which a peat body can be identified as a single hydrological entity (though, in the case of blanket bog mesotopes, these may have hydrological links with other mesotopes), e.g. Cors Fochno (Borth Bog) in central Wales or Brishie Bog in the Silver Flowe. The lagg fen around a raised bog is a distinct mesotope, with its own hydrological requirements, so a complete raised bog system, with its lagg fen, should be classed as a macrotope.

3.1.3　　The **mire margin/mire expanse gradient** (Sjors 1948). The mire expanse is the main bog surface area which may support distinctive and repeated surface patterning (microtope: see section 3.1.4 below); the mire margin is generally simpler in surface structure and represents areas of thinner peat. The mire margin may actually form the edge of the mire, as in the case of a raised bog rand (see section 1.3), or it may simply represent the thinner peat between areas of deeper, patterned ombrotrophic peat in blanket bog (Figure 2). Both the animal and the plant communities of the mire margin differ markedly from those of the mire expanse. Its relative ecological simplicity should not be mistaken for lack of importance; both the margin and the lagg fen play fundamental parts in the hydrological integrity of a bog (Lindsay in press).

3.1.4　　Mire **microtope**, relating to the arrangement of surface features, especially into a pattern which alternates aquatic and terrestrial elements, e.g. pool and hummock, or terrestrial features alone, e.g. hollow and ridge (Lindsay in press).

3.1.5　　Mire **microform**, relating to single surface features, such as pool or hummock. (See also Figure 5 and Lindsay *et al.* 1988, pp. 23-24).

3.2　　Within mire macrotopes and mesotopes are different varieties of raised bog and blanket bog, the representation of which will be dealt with in detail below. The National Vegetation Classification (Rodwell 1991) is perhaps better suited to the general description of the mire margin/mire expanse gradient. In intact bogs, it may be possible to assign the mire margin to a single National Vegetation Classification type, and the mire expanse to a larger number of types. Cut-over and drained mires are usually more complex to classify.

3.3　　The other structural levels (microtope, microform) can be addressed separately. Within the mire microtope, ten kinds of microform have been recognised (see Table 3, Lindsay *et al.* 1985, and Lindsay *et al.* 1988). These are related to vegetation types (Figure 3), resulting from competitive adaptations to water table. On any given site, microtope features will usually occur in a small number of repeating combinations, typically no more than three. There will also be a small range of vegetation types within these microtopes. The National Vegetation Classification gives a valuable summary of mire plant communities at the level of general ombrotrophic mire vegetation and for the broad distinctions between mire expanse and mire margin or between aquatic phase and terrestrial phase in microtope patterns. Its vegetation types are not sufficiently detailed to describe the intricate fine-scale vegetational mosaics of patterned surfaces which vary across Britain. For these a further subdivision of the National Vegetation Classification categories has been devised (Lindsay in press). The National Vegetation Classification treatment for ombrotrophic mires is given in Table 1 and the finer classification in Figure 3.

3.4　　**Within each AOS, selection of sites should aim to represent the range of variation in each of the five structural levels (3.1.1 - 3.1.5) that are present,**

and the range of plant communities and sub-communities associated with mire microforms, bearing in mind local and regional climatic variations. However, many AOSs may have only one or two of the structural types present.

To help prioritise sites above the minimum standards of size and peat formation capability (section 3.5), there are certain general features which indicate the most natural sites, which may be assumed to have the greatest quality.

3.4.1   parts of the original lagg fen still present (in raised bogs particularly);

3.4.2   in raised bogs, a high proportion of the original central dome still physically intact;

3.4.3   low frequency of drains and peat-cuttings;

3.4.4   presence of plant species indicating peat formation capability and/or lack of disturbance, notably *Sphagnum pulchrum, S. fuscum, S. imbricatum, S. balticum, S. magellanicum, Dicranum bergeri, Rhynchospora alba, R. fusca, Drosera anglica, Carex limosa, C. magellanica* (= *C. paupercula*) and locally *Schoenus nigricans*. Other component species are described in Annex 1.

3.4.5   an area of natural surface pattern (as defined in 3.1.4) within the mire expanse; and

3.4.6   absence of invasion by woodland or scrub, though some high quality sites may contain trees and scrub with a bog bryophyte floor.

There should be a presumption towards selection of any site exhibiting two or more of these characteristics. Furthermore, this presumption should apply to all sites above the minimum standards of size and peat formation capability (section 3.5).

3.5   The general quality of a bog is assessed by the degree to which it has remained intact as a hydrological and vegetational system capable of active peat growth. **Peat formation capability** for both raised and blanket bogs is defined by the hydrological and biotic features described in Annex 1, the definition of 'active' raised and blanket bogs adopted by the European Union.

Raised bogs larger than 10 ha and blanket bogs larger than 25 ha should be considered for SSSI status in all parts of Britain if capable of forming peat. Smaller raised bog sites of high quality may be selected in Areas of Search where few or no larger sites remain.

Some bogs naturally have a low cover of *Sphagnum* (e.g. those in eastern England) or have remained substantially intact despite an increase in dominance of vascular plants with a slower capacity for peat formation. Erosional features also have considerable scientific interest (Lindsay in press) and need to be included in assessment and selection (see Table 3 and Figure 3).

4. Raised bogs

4.1 Formerly an extensive habitat in waterlogged areas in lowland Britain, raised bog in near-natural condition has been reduced to less than 4,000 ha, and almost all remnants are modified.

There should therefore be a presumption towards the selection of all examples above the minimum standard (largely measured by size and capability for peat formation (see section 3.5)).

Cut-over, drained or afforested raised bogs may be selected (see 4.2), where, for example, some of the surface is still uncut or natural regeneration is already evident, but also in AOSs where no intact examples remain. Furthermore, in AOSs where all examples are severely damaged, the best examples may be selected, especially where rare or protected species are present.

The following mesotopes should be assessed separately:

4.1.1 *Flood-plain raised bog*

These bogs are discrete areas bounded by mineral ground, usually at low levels on alluvial or fluvioglacial flood plains. In most cases the surrounding flood-plain fens have been reclaimed for agriculture. A few examples occur in upland situations.

There should be a presumption towards the selection of all examples which support more than 10 ha of bog vegetation with peat formation capability, even where there has been modification of the original bog margin and expanse features. This presumption should also apply if any one or more of these microtopes is present:

- **Ridges of *Sphagnum - Rhynchospora alba, Sphagnum - Andromeda* or *Sphagnum pulchrum* (T1, see Table 3)**

- **Bryophyte hummocks of *Sphagnum fuscum, S. imbricatum* or, more typically, *S. capillifolium* and *S. magellanicum* (T3/2)**

- **Hollows of *Sphagnum - Rhynchospora alba* or *Sphagnum - Drosera anglica* (A1)**

Furthermore there should be a presumption towards the selection of all examples retaining more than 50% of their peripheral lagg systems. In AOSs where few intact sites exist, cut-over examples may be selected subject to the conditions set out in section 4.2.

4.1.2   *Estuarine raised mire*

This type is formed over estuarine sediments resulting from ancient land/sea level changes. Where transitions to maritime habitats still survive, it is important to ensure that these are included in the site. Both this present relationship with the marine environment and the stratigraphic sequence revealing historical connections are important attributes, but these may not become clear until coring surveys have been carried out.

The guidelines under section 4.1 therefore apply, but degraded surfaces may be selected if the stratigraphy is deemed sufficiently important.   Expert   opinion   on   the   palaeoenvironmental significance should be sought.

4.1.3   *Basin raised bog*

These bogs are usually associated with, and grade into, basin fens (Chapter 7, section 4.2, NCC 1989). There has usually been a seral development through open water and fen phases, and the fen may persist only as a lagg around the central peat dome, though it may occur as a less regular mosaic within the basin. In some sites fen is now lacking altogether.

The criteria under section 4.1 apply to most basin raised mire sites with the exception of those examples which have developed as *Schwingmoor* in kettleholes (see Chapter 7, Figure 3, NCC 1989) where the water surface has been invaded by vegetation, subsequently closing to form a peat dome. This is a rare type, and there should be a presumption towards the selection of all such sites, unless they are severely degraded (including by pollution) beyond recovery or are very small (less than 1 ha).

For palaeoecological studies, it is important to ensure that a full range of variation in stratigraphic sequences (including their geographical variation) is represented. An assessment of the stratigraphy may not be possible without detailed coring surveys of the sites.

4.2   Damaged raised bog

Provided such sites retain significant areas of peat-forming vegetation (see Annex 1. para 2), there are two circumstances where it may be necessary to include damaged bog within an SSSI.

First, in AOSs which retain only damaged raised bog (e.g. completely cut-over sites) such sites may be selected even if the damage is considered severe. Many such sites retain great intrinsic biological and scientific interest.

Second, where a remnant of primary dome (Lindsay in press) survives, surrounded by a damaged surface, inclusion of some or all of the damaged area within the site will usually be necessary to protect the identified biological and scientific interest (see section 7, Boundary selection).

In the case of commercial peat extraction, cutting may be continuing because of extant planning consent. The conservation priority should nevertheless be to secure mire regeneration as an accepted after-use. Perched remnants of peat dome tend to exhibit a narrower range of species as they dry out, but hydrological restoration may be possible and can contribute to the conservation management of the adjoining cut-over areas.

In the medium term, the recovery of damaged areas is only possible if a suitable hydrological regime can be provided. Opportunities for the expansion of actively growing bog habitat are therefore restricted to the present extent of the bog peat soils. General criteria which should influence selection are the maintenance of suitable water levels and water quality and a minimum depth of residual peat on cut-over areas. Absence of colonising woodlands is also considered an advantage. Carefully monitored reintroductions might be considered but only from local donor sites.

Whilst the stratigraphic/pollen record cannot be replaced, given optimum conditions for regeneration, the natural vegetation pattern for the climatic zone will almost certainly develop over a very long period (centuries, not decades).

In AOSs where there are few, if any, intact sites, then cut-over, drained or afforested sites, which exhibit none of the characteristics listed in 3.4.1 - 3.4.6, may be selected if the following two criteria are met:

- **a significant area of vegetation with peat formation capability is present within the site;**

  and

- **the hydrology has not been irreparably damaged.**

5. Intermediate mires

5.1 These mires occur under conditions of climate and topography which are marginal for the development of blanket bog (Lindsay in press). Stratigraphic surveys are necessary to determine their developmental history, but many examples appear to be bogs which have expanded laterally to engulf any original rand and lagg features, so that the edge of the peat mass tends to merge gradually into surrounding areas of mineral soil through decreasing depth of peat. Sometimes two adjacent peat lenses have coalesced across a low intervening mineral ridge to form a single mire expanse. Intermediate mires tend to be found at low elevations (30-150 m) and so often occur as isolated expanses of peat within a non-peat landscape, with artificial margins resulting from reclamation for agriculture or peat-cutting. Peat depths may exceed 4 m in the deepest parts and there may also be a slightly domed appearance as in raised bogs. Intermediate mires belong especially to the far north of England (north Cumbria and Northumberland), southern Scotland and the Central Lowlands. They may occur as scattered and isolated peat masses but where extensive, they may grade into true blanket bog.

For the purposes of SSSI selection, intermediate mires should be classified either as raised mire if they have indications of a rand/lagg system, or, more typically, as low altitude blanket mire. They may be selected as representing part of the range of variation between the two main mire types. Their intermediate character should be mentioned in the description of their scientific interest.

6. Blanket bog

6.1   Despite extensive past degradation, reclamation for agriculture and recent afforestation, blanket bog is still one of the more extensive natural or semi-natural formations remaining in Britain. This gives grounds for regarding it as a type which should be represented by the selection of exemplary sites showing the full range of ecological variation. A more significant factor still is the international importance of blanket bog in Britain. Blanket bog is recognised as a globally rare formation (see section 2.2). Site selection should pay special regard to this international dimension, including the need to meet international commitments (see part B, section 3, NCC 1989). See also Section 6.6 of this document for guidance relating to the Flow Country.

6.2   From the manner of its development, blanket bog often represents a complex of mire units (mesotopes) and typically includes minerotrophic (especially soligenous) elements and transitions to vegetation on non-peat soils. Usually, a range of fairly distinct mesotopes can be identified within the general expanse of more or less peat-covered land (Lindsay in press). These should be examined for hydrological connections, and linked groups then drawn together into macrotopes. However, for selection purposes, individual mesotopes should be described and classified as separate units first; then the macrotopes should be classified on the basis of their combinations of individual mesotopes. Sometimes mesotope boundaries are particularly indistinct, especially where the bog has a relatively smooth surface configuration giving a fairly uniform appearance, and definition may therefore be difficult. In such cases it is preferable to recognise a mesotope complex (macrotope) and to treat this as the unit of assessment and selection. Component mesotopes are watershed, valleyside, spur, and saddle mires (among bog types) with soligenous and topogenous fen units (see Chapter 7, sections 4-5, NCC 1989). Non-peatland features may include wet heath and grassland; dry heath and grassland; streams and rivers; lochs, lakes and tarns; and rock outcrops. Because blanket bogs usually merge imperceptibly into other natural or semi-natural habitats within upland areas, it is often appropriate to regard them as part of a still larger ecosystem complex, which then becomes the unit of assessment (see Chapter 9, section 4, NCC 1989).

6.3   The major mire mesotopes (or units) representing the 'building blocks' within blanket bog complexes are described below and illustrated in Figure 6. Where pool and hummock systems occur, these units often appear well-defined but where this juxtaposition is absent, they may be less easily recognisable.

6.4   Blanket bog mesotopes which may be present in the macrotope are:

6.4.1   *Watershed mire*

This type occurs on watershed plateaux or broad ridges, where the surrounding land slopes away on all sides. The topographic situation is varied, from flat hill tops with steep slopes below (e.g. Kinder Scout in the Peak District) to the gently contoured moorlands of Caithness and east Sutherland characterised by extensive 'flows'. This is the most evidently 'ombrotrophic' mire type. There is no higher ground from which ground-water could drain, so that the only source of water is certainly precipitation. In England and Wales, where surface patterns occur, they are mostly only small *Sphagnum* hollows or scattered pools. In the north and west, watershed bogs show a wide range of surface patterning, and in the far north of Scotland they exhibit extreme aquatic features with large, deep and rounded open-water pools (*dubh lochain*).

6.4.2   *Valleyside mire*

This type occurs on gently sloping or almost level ground lying between higher, steeper terrain and a watercourse which forms its lower margin. Slight enrichment may be apparent in the vegetation, for example increased *Molinia*, and, where there is frequent or strong water movement through a peat surface, these areas may grade into sloping fen. Valleyside mires are most clearly defined where there is surface patterning, typically with curved elongated pools following the contours. Aquatic zones include hollows (Al-A2) and pools (A3) (see Table 3), and where deep pools (A4) occur, they are typically near the upslope limit of the system. Whilst watershed and valleyside mires appear distinctively different when separated by a break in slope, there are many situations where one grades into the other, making distinction between them almost impossible. Thus a single mound of mineral ground, bounded on all sides by streams, may be cloaked by a mound of peat which is both watershed and valleyside mire. Such systems are often referred to as watershed/valleyside, to distinguish them from hill top watershed mires.

6.4.3   *Spur mire*

Where the shoulder of a hill flattens into a broad spur, its crest often carries a lobe of blanket bog with a distinctive form. Some of the bog will have the character of watershed mire, but ground closest to the slope above the spur may receive drainage water, providing a minerotrophic or soligenous influence. Occasionally, the spur topography has a basin-like form, and this may give rise to a peat lens resembling that of raised bog. The distinctive feature separating this from valleyside mire is that its lower edge is not associated with a river or lake but instead is delimited by a steepening slope.

6.4.4   *Saddle mire*

This is similar in many respects to spur mire, but it lies in the depression between two higher slopes and so may receive a soligenous influence at each end. The mire may be largely ombrotrophic if the higher ground at each side slopes gradually upwards from the col. Depending on the angles of slope below the saddle, the mire extends downwards on either side, giving the appearance of a horse's saddle.

6.4.5   *Eccentric mire*

Considered by some as a specific bog type, eccentric mire has not been widely recorded in Britain. Moore (1977) describes Claish Moss, in Argyll, as an eccentric mire complex. Lindsay (in press) observes that Claish, Kentra Moss and possibly Blar na Caillich Buidhe, all occur within a limited region of Lochaber and Argyll. Davis & Anderson (1991) provide a detailed account of this ombrotrophic mire type for North America as do Sjörs (1983) and Moen (1985) for Fennoscandia. Eccentric bogs slope mainly in one direction and occur on valleysides. They abut mineral ground at their upslope margin, whereas the downslope limit borders on an unpatterned fen which in turn may border a lake or a stream in the valley floor. The bog slope normally has a surface pattern of near-linear ridges and hollows aligned at right angles to the slope. The whole bog can be fan-shaped, with the narrow part upslope. Surface features are characterised by hummocks, ridges, hollows and pools, all arranged in a distinctly arcuate microtope aligned at right angles to the direction of water seepage. The patterns are very much more extreme than is typical for normal valleyside mire. Erosion hags and gullies may also be present. Peat mounds are not recorded from eccentric bogs.

6.5   General representation of blanket bog

In addition to applying the general principle of representing the **range of variation** within each AOS, it is important to ensure that an adequate **extent** of blanket bog is included in the series of selected sites. Selection according to the concept of the **topographic unit,** described under Upland habitats (Chapter 9, sections 4.3 and 4.7, NCC 1989), should be applied in upland areas where blanket bog is extensive, so that not only structural and vegetational features, but also dependent animal assemblages, are well represented. In some areas blanket bog is only one element, though a major one, in a complex of features which together form an important landscape unit, and it is necessary to include associated streams, lakes, drier terrain and rock outcrops. Sometimes, patches of blanket bog (and fen) are scattered widely over an area of upland which consists predominantly of other habitats and it is appropriate to select a boundary which contains an adequate series of these separate patches.

Patterned bogs with peat formation capability are of particular interest, and there should thus be a presumption towards the selection of any such site larger than 25 ha or which shows unusual microtopic features. This presumption towards selection should also apply to all examples of peat mound (T5) 'fields' over 10 ha.

Blanket bog mesotopes showing any of the following microtopic and vegetational features are near-natural and of high quality. Subject to the minimum standards of size and peat formation capability set out above, there should be a presumption towards the selection of sites which contain:

- **An abundance of *Sphagnum*-rich ridges (T1)**

- **An abundance of *Sphagnum*-rich ridges (T2)**

- **Ridges of *Sphagnum - Betula nana* (T2)**

- **Bryophyte hummocks of *Sphagnum fuscum* or *S. imbricatum* (T3/2)**

- **Peat mounds (T5)**

- **Hollows of *Sphagnum* or bare peat - *Rhynchospora fusca* (A2)**

6.6 The Flow Country

The enormous system of blanket bogs occupying much of Caithness and east Sutherland, with more fragmented outliers in west Sutherland, has been highlighted as a particularly important case for SSSI designation. This, the Flow Country, is one of the most important systems of blanket bog in the world and regarded by international opinion as a candidate for listing under the World Heritage Convention (IUCN 1988). It is important that the SSSI series chosen for the Flow Country adequately represents both the global importance of the area and the full range of peatland variation within it.

7. Boundary selection for raised and blanket bogs

7.1 The identification of ground and water which provide the continued long-term support of the hydrological and ecological functioning of a bog system underpin the rationale for boundary selection. So, bogs must be protected at their margins from potentially damaging activities, especially those activities likely to cause hydrological disturbance by maintaining or increasing water run-off by artificial drainage.

Site boundaries must be chosen to include all land judged necessary to provide and maintain the hydrological functions needed to conserve the special features of the site.

7.2   Bog systems subject to drainage influences from higher ground need to be protected by the inclusion of an adequate catchment buffer zone in the same ways as fens (Chapter 7, section 9, NCC 1989). This applies not only to small and isolated basin raised bogs, but to raised bogs in mountain valleys and various types of blanket bog other than watershed mires. In many situations there will be a risk of future afforestation or other disturbance upslope, with the ensuing probability of changes in hydrology and in the chemical content of drainage water from the catchment, altering the input to the mire system.

7.3   Where a bog system is bounded by agricultural land, the site boundary may need to follow the original extent of the peat body. In many cases this may mean that the boundary includes some agricultural land on peat, **but only if it still plays a functional part in the overall hydrology of the peat body containing the special interest.** The purpose of including such land is to prevent an increase in the hydraulic gradient at the bog margin which can occur as a consequence of water losses by drainage which lead to oxidation, slumping and peat shrinkage. Land-uses which do not require further drainage of this ground may be compatible with the maintenance of its functional role.

7.4   If part of a raised bog has been cut over or afforested, it may still be important to include this area as an integral part of the system (see also 4.2). If lagg systems are present, boundaries should not run along their centre (or stream) but along their outer limit bounded by mineral ground. Because of the need to apply the topographic unit principle (see 6.5), the boundaries of blanket bog SSSIs may have to be drawn well outside the peatland edge.

In some circumstances it may be necessary to seek expert hydrological advice prior to deciding the boundary of a candidate SSSI.

SUMMARY OF SELECTION PROCESS

STAGE 1   Classify sites within AOS into general bog types - raised bog (go to A) or blanket bog (go to B).

A - RAISED BOG

STAGE 2   Using Section 4.1, determine hydromorphological types - flood-plain, estuarine or basin.

STAGE 3   Determine microtope and microform features within sites (especially on the mire expanse), looking in particular for **surface patterning** as an indicator of the highest quality sites. Classify sites according to the full range of such features within the AOS.

STAGE 4   Determine biotic features of sites to ensure:

    a)   the full range of vegetation on mire expanses and margins within AOS is represented; and

    b)   determine the presence and extent of vegetation capable of peat formation; and

    c)   ensure that the habitats of rare and specialised fauna (particularly invertebrates and birds) are protected.

STAGE 5   Apply 'minimum standard' criteria in Sections 3 and 4 to identify sites with a presumption for selection under these criteria. In AOS where all sites are severely degraded then the best examples may be selected.

STAGE 6   Determine boundaries of selected sites using guidelines in Section 7.

B - BLANKET BOG

STAGE 2   Using Section 6.4, as far as possible determine hydromorphological types - watershed; valleyside; spur; saddle; or eccentric.

STAGE 3   Identify sites within AOS where there are areas of vegetation over 25 ha showing peat formation capability.

STAGE 4   Apply criteria in Sections 3 and 6 to select the best sites to represent the full range of microtopes, microforms and vegetational and faunistic types present within the AOS.

STAGE 5   Determine site boundaries using the guidelines in Section 7.

8.   References

Barber, K.E. 1981. *Peat stratigraphy and climate change*. Rotterdam, Baekema.

Birks, H.H. 1975. Studies in the vegetational history of Scotland. IV. Pine stumps in Scottish blanket peats. *Philosophical Transactions of the Royal Society of London, 270B*: 181-226.

Birks, H.J.B. 1988. Long-term ecological change in the British uplands. *In: Ecological change in the uplands*, ed. by M.B. Usher & D.B.A. Thompson, 37-56. Oxford, Blackwell Scientific Publications.

Bridge, M.C., Haggart, B.A., & Lowe, J.J. 1990. The history and palaeoclimatic significance of subfossil remains of *Pinus sylvestris* in blanket peats from Scotland. *Journal of Ecology, 78*: 77-99.

Davis, R.B., & Anderson, D.S. 1991. *The eccentric bogs of Maine: a rare wetland type in the United States*. Maine, University of Maine. (Maine State Planning Office, Critical Areas Programme, Planning Report 93.)

Gear, A.J., & Huntly, B. 1991. Rapid changes in the range limits of Scots Pine 4,000 years ago. *Science, 251*: 544-547.

Godwin, H. 1975. *The history of the British flora. A factual basis for phytogeography*. 2nd ed. Cambridge, Cambridge University Press.

Grünig, A., Vetterli, L., & Wildi, O. 1986. *Les hautes-marais et marais de transition de Suisse: resultats d'un inventoire*. Birmensdorf, Swiss Federal Institute of Forestry Research.

International Union for Conservation of Nature and Natural Resources. 1988. *17th Session of the General Assembly of IUCN: San José, Costa Rica, 1-10 February 1988. Resolutions and recommendations*. Gland, Switzerland.

Ivanov, K.E. 1981. *Water movement in mirelands*. London, Academic Press. (Translated from Russian by A. Thomson and H.A.P. Ingram.)

Lindsay, R.A. In press. *Bogs: the classification, ecology and conservation of ombrotrophic mires*. Edinburgh, Scottish Natural Heritage.

Lindsay, R.A., Charman, D.J., Everingham, F., O'Reilly, R.M., Palmer, M.A., Rowell, T.A., & Stroud, D.A. 1988. *The Flow Country. The peatlands of Caithness and Sutherland*, ed. by D.A. Ratcliffe and P.H. Oswald. Peterborough, Nature Conservancy Council.

Lindsay, R.A., Riggall, J., & Burd, F.H. 1985. The use of small-scale surface patterns in the classification of British peatlands. *Aquilo, Seria Botanica, 21*: 69-79.

Moen, A. 1985. Classification of mires for conservation purposes in Norway. *Aquilo, Seria Botanica, 21*: 95-100.

Moore, P.D. 1977. Stratigraphy and pollen analysis of Claish Moss, north-west Scotland: significance for the origin of surface pools and forest history. *Journal of Ecology, 65*: 375-397.

Nature Conservancy Council. 1989. *Guidelines for selection of biological SSSIs.* Peterborough.

Rodwell, J.S., *ed.* 1991. *British plant communities. Vol. 2: Mires and heaths.* Cambridge, Cambridge University Press.

Sjörs, H. 1948. Myrvegetation i Bergslagen. [Mire vegetation in Bergslagen, Sweden]. *Acta Phytogeographica Suecica, 21*: 1-299. [English summary: 277-299.]

Sjörs, H. 1983. Mires of Sweden. *In: Mires: swamp, bog, fen and moor. Regional studies*, ed. by A.J.P. Gore, 69-94. Amsterdam, Elsevier Scientific. (Ecosystems of the World, 4B.)

Steiner, G.M. 1992. *Osterreichischer Moorschutzkatalog.* [Austrian mire conservation catalogue.] Grüne Reihe des Bundesministeriums für Umwelt, Jungend und Familie, Band 1. Graz, Ulrich Moser.

Stevenson, A.C., Jones, V.J., & Battarbee, R.W. 1990. The cause of peat erosion: a palaeolimnological approach. *New Phytologist, 114*: 727-735.

Stroud, D.A., Mudge, G.P., & Pienkowski, M.W. 1990. *Protecting internationally important bird sites.* Peterborough, Nature Conservancy Council.

Tubridy, M. 1984. *Creation and management of a Heritage Zone at Clonmacnoise, Co. Offaly, Ireland.* Dublin, Trinity College.

Table 1 National Vegetation Classification - bog communities and subdivisions (Rodwell 1991) and bog microtopes and communities (Lindsay *et al*. 1985; Lindsay *et al*. 1988)

1   Mire margin and smooth blanket mire

NVC	category
M15	*Scirpus cespitosus - Erica tetralix* wet heath
M15a	*Carex panicea* sub-community
M15b	Typical sub-community
MI5C	*Cladonia* spp. sub-community
M15d	*Vaccinium myrtillus* sub-community
M16	*Erica tetralix - Sphagnum compactum* wet heath
M16a	Typical sub-community
M16b	*Succisa pratensis - Carex panicea* sub-community
M16d	*Juncus squarrosus - Dicranum scoparium* sub-community
M17	*Scirpus cespitosus - Eriophorum vaginatum* mire
M17b	*Cladonia* spp. sub-community
M17c	*Juncus squarrosus - Rhytidiadelphus loreus* sub-community
M18	*Erica tetralix - Sphagnum papillosum* raised and blanket mire
M18a	*Sphagnum magellanicum - Andromeda polifolia* sub-community
M18b	*Empetrum nigrum* ssp. *nigrum - Cladonia* spp. sub-community
M19	*Calluna vulgaris - Eriophorum vaginatum* blanket mire
M19a	*Erica tetralix* sub-community
M19b	*Empetrum nigrum* ssp. *nigrum* sub-community
M19c	*Vaccinium vitis-idaea - Hylocomium splendens* sub-community
MI9ci	*Betula nana* variant
MI9cii	Typical variant
M19ciii	*Vaccinium uliginosum* variant
M20	*Eriophorum vaginatum* blanket and raised mire
M20a	Species-poor sub-community
M20b	*Calluna vulgaris - Cladonia* spp. sub-community
M21	*Narthecium ossifragum - Sphagnum papillosum*
M21b	*Vaccinium oxycoccos - Sphagnum recurvum* sub-community
M2	*Sphagnum cuspidatum/recurvum* bog pool community
M2b	*Sphagnum recurvum* sub-community

2   Mire expanse

2.1 Hummock/high ridge/erosion hag (T2-T4)

M17	*Scirpus cespitosus - Eriophorum vaginatum* mire
M17b	*Cladonia* spp. sub-community
M18	*Erica tetralix - Sphagnum papillosum* raised and blanket mire
M18a	*Sphagnum magellanicum - Andromeda polifolia* sub-community
M18b	*Empetrum nigrum* ssp. *nigrum - Cladonia* spp. sub-community
M19	*Calluna vulgaris - Eriophorum vaginatum* blanket mire
M19a	*Erica tetralix* sub-community
M19b	*Empetrum nigrum* ssp. *nigrum* sub-community
M19ci	*Betula nana* variant

2.2 Low ridge (T1)

M2	*Sphagnum cuspidatum/recurvum* bog pool community
M2a	*Rhynchospora alba* sub-community

M15b	*Scirpus cespitosus-Erica tetralix* wet heath, *typical* sub-community
M16c	*Ericetum tetralicis* wet heath, *Rhynchospora alba-Drosera intermedia* sub-community
M17	*Scirpus cespitosus - Eriophorum vaginatum* mire
M18a	*Sphagnum magellanicum - Andromeda polifolia* sub-community

2.3 Hollows (A1, A2)

M1	*Sphagnum auriculatum* bog pool community
M2	*Sphagnum cuspidatum/recurvum* bog pool community
M3	*Eriophorum angustifolium* community
M16c	*Ericetum tetralicis* wet heath, *Rhynchospora alba-Drosera intermedia* sub-community

2.4 Pools (A3, A4)

M1	*Sphagnum auriculatum* bog pool community
M3	*Eriophorum angustifolium* community
M4	*Carex rostrata-Sphagnum recurvum* mire
M16c	*Ericetum tetralicis* wet heath, *Rhynchospora alba-Drosera intermedia* sub-community

Table 2 Present day remaining area of near-natural bog vegetation as a proportion of the area of land under soils formed from raised bogs

	Amount of land with raised bog soils		Amount of land with near-natural bog vegetation			
	Area (ha)	No. of Sites	Area (ha)	(*)	No. of Sites	(*)
England	37,413	210	493	(1)	15	(7)
Scotland	27,892	851	2,515	(9)	129	(15)
Wales	4,086	21	818	(20)	6	(29)
	69,391	1,082	3,826	(6)	150	(14)

(*) values in parentheses are percentages

Table 3 Mire microforms

Terrestrial (T) zones

(T1)   Low ridge ("lawn": Sjors 1948) - common on mire areas which are free from damage; 1-10 cm above the mean water table; generally the richest zone for characteristic mire species.

(T2)   High ridge - the general level of many mire surfaces, particularly outside pool systems; 10-20 cm above the mean water table.

(T3)   Hummock - normally the highest element in the pattern and always bryophyte-formed; 20 cm to 1 m above the mean water table.

(T4)   Peat hag - associated with erosion; 1-2 m above the mean water table.

(T5)   Peat mound - occurs only in Shetland, Caithness, Sutherland and the Outer Hebrides; 1-3 m above the water table and possibly linked to incipient 'palsa' form, though the origins are as yet obscure.

Aquatic (A) zones

(A1)   *Sphagnum* hollow ("carpet": Sjors 1948) - a true hollow (i.e. aquatic phase) of dense *Sphagnum cuspidatum*; 0-10 cm below the mean water table.

(A2)   Mud-bottom hollow (Sjors 1948) - a hollow dominated by a relatively solid bare peat base, but with some aquatic *Sphagna*; 5-20 cm below the mean water table; not recorded from eastern Britain (including Caithness).

(A3)   Drought-sensitive pool (Lindsay *et al.* 1988) - an area of open water with an unconsolidated peat base which remains flooded for much of the time but in drought conditions will dry up; 20-50 cm below the mean water table.

(A4)   Permanent pool (Lindsay *et al.* 1988) ("summer pool": Tubridy 1984) - an area of open water which is sufficiently deep to remain flooded even during extreme drought; 1-4 m deep; restricted to north-west Strathclyde, Tayside and regions north of them.

(TA2) Erosion gullies, resembling mud-bottom hollows but with flowing water.

These surface features are arranged into patterned areas in various combinations (see Figure 5). The range of surface patterns contributes significantly to variability within and between sites. **This range should therefore be represented in the selection process.** The distribution and abundance of particular levels or zones in areas of patterning provide one level of selection, but in addition the form and orientation taken up by the patterns are an important factor. Lindsay *et al.* (1985) indicated in general terms the geographical variation displayed by these patterns across Britain. A site may, for example, consist purely of low ridge (T1) and high ridge (T2) without any true aquatic phase. Increasing wetness of climate gives rise to patterned areas of increasing complexity. In the driest areas of bog formation in Britain the aquatic phase, if it exists at all, tends to form small unaligned hollows (A1/A2), but with increasing wetness these hollows become markedly linear. Open water hollows (A3) demonstrate extreme linear patterning towards the north and west of Scotland, whilst open-water pools (A4) are characteristically rounded, formed on the top of watersheds and restricted to the most northerly oceanic areas of Britain.

Erosion patterns and features can also be important characters in comparisons of mire mesotopes and macrotopes. The most obvious features are the deep erosion gullies and hags typical of many plateau and watershed sites. Further north, erosion features include empty pools, leaving exposed beds of peat or even bedrock. Deep gully erosion is a well-known feature of peat in the Pennines, with gullies attaining depths of 2-3 m. However, if an erosion complex forms only in the surface skin of peat comprising the top few centimetres, both the gullies and the hags tend to be extremely small, with hags no more than 20-25 cm high and with diameters of 10-30 cm, surrounded by a network of interconnecting shallow channels. This is not intense erosion, as many channels support a wet matrix of *Sphagnum* and peat; nor, however, is it completely intact mire. The term 'microbroken' has been coined to classify this particularly abundant mire feature. On aerial photographs the mire surface appears to be dimpled or covered with a dense mass of rounded papillae, rather than with the dramatic linear patterns or heavy reticulate networks associated with hag and gully erosion. This stage may later develop into more serious gullying or sheet erosion.

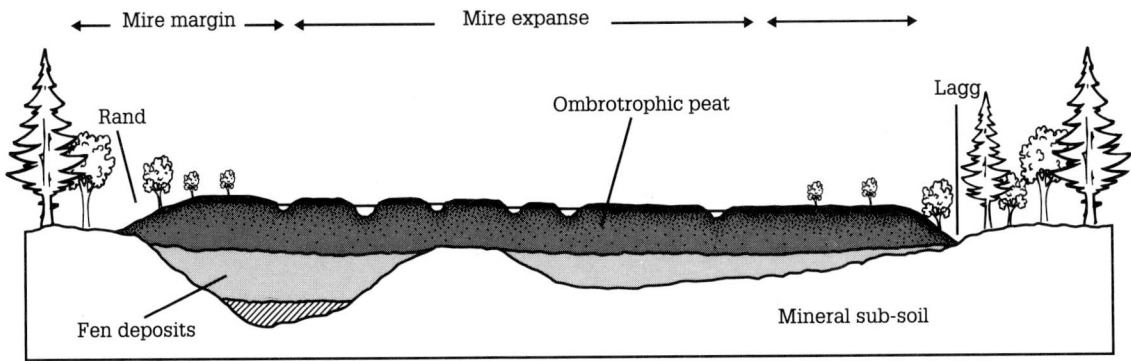

Figure 1 Profile of a plateau ("continental") raised bog. Oceanic raised bog is typically more domed throughout its profile and has less tree cover, but otherwise the components making up the profile in the two types are similar. (Adapted from Grünig *et al.* 1986).

Supplement to *Guidelines for selection of biological SSSIs* (Nature Conservancy Council 1989)

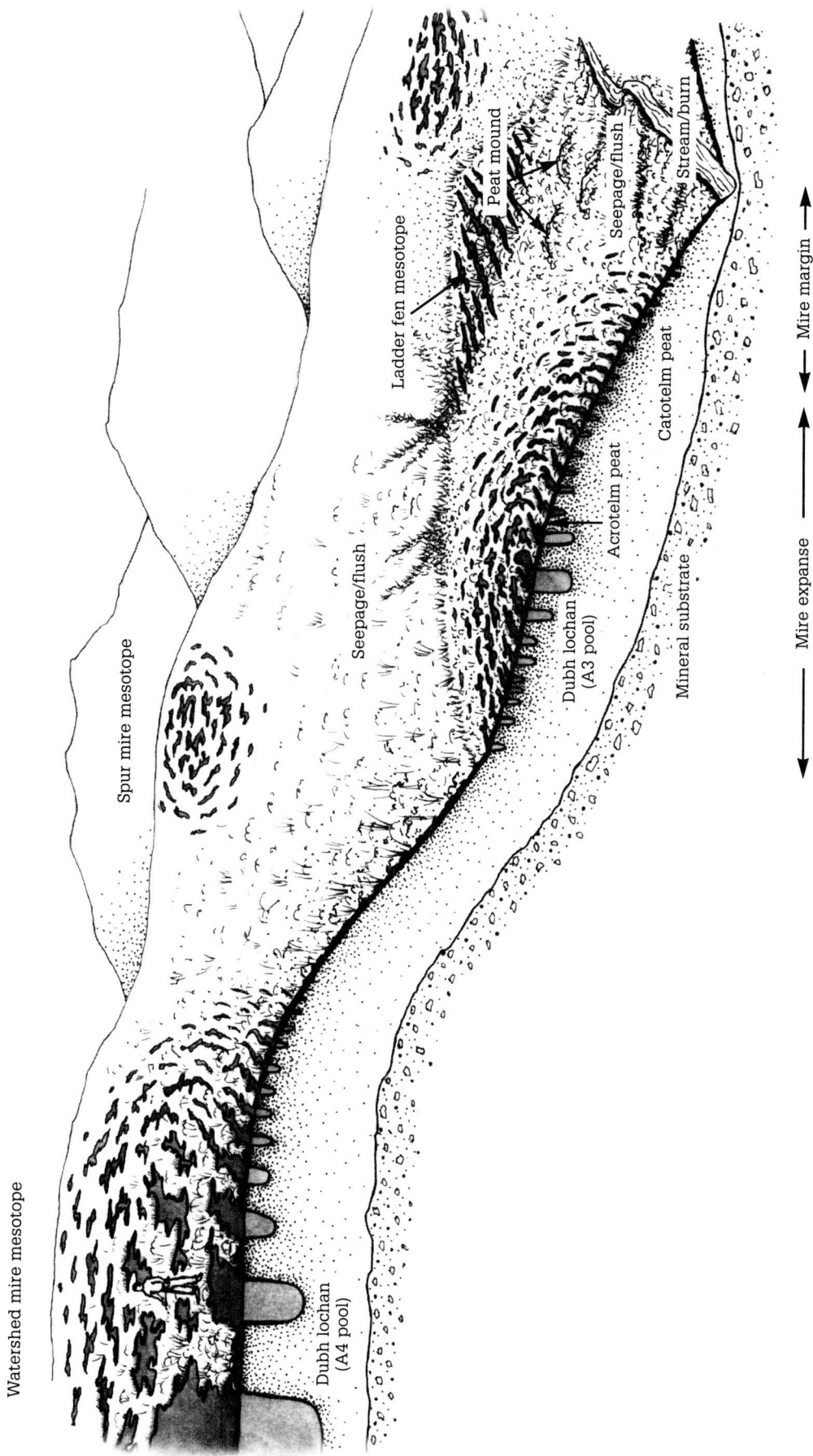

Figure 2   Characteristics of a blanket mire complex from one of the wettest parts of Britain (e.g. Sutherland), showing extreme forms of pool-and-ridge patterning.   (Adapted from Lindsay *et al.* 1988).

Supplement to *Guidelines for selection of biological SSSIs* (Nature Conservancy Council 1989)

Terrestrial (T)

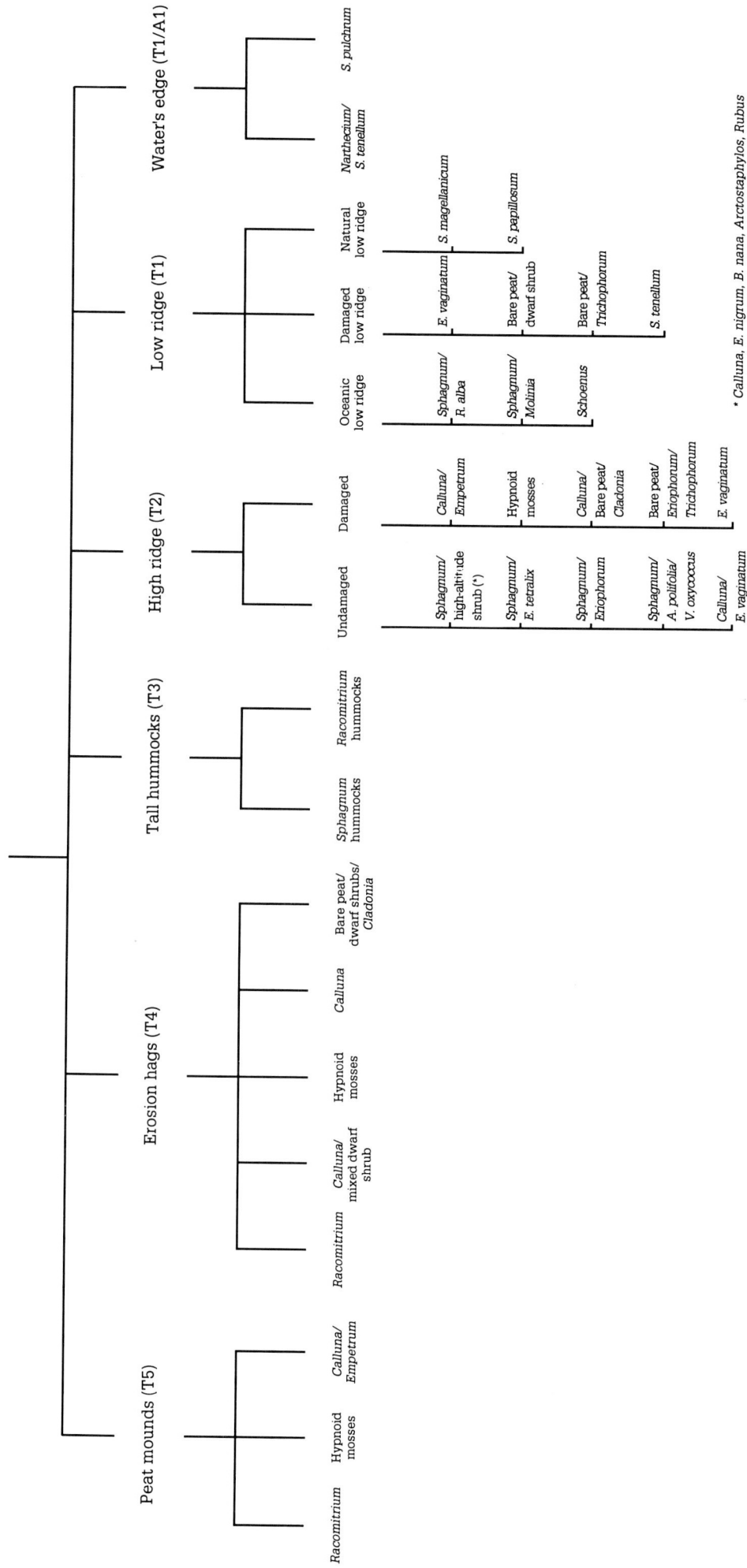

Peat mounds (T5) — Erosion hags (T4) — Tall hummocks (T3) — High ridge (T2) — Low ridge (T1) — Water's edge (T1/A1)

Peat mounds (T5):
- *Racomitrium*
- Hypnoid mosses
- *Calluna/ Empetrum*

Erosion hags (T4):
- *Racomitrium*
- *Calluna/ mixed dwarf shrub*
- Hypnoid mosses
- *Calluna*
- Bare peat/ dwarf shrubs/ *Cladonia*

Tall hummocks (T3):
- *Sphagnum* hummocks
- *Racomitrium* hummocks

High ridge (T2):

Undamaged:
- *Sphagnum/ high-altitude shrub* (*)
- *Sphagnum/ E. tetralix*
- *Sphagnum/ Eriophorum*
- *Sphagnum/ A. polifolia/ V. oxycoccus*
- *Calluna/ E. vaginatum*

Damaged:
- *Calluna/ Empetrum*
- Hypnoid mosses
- *Calluna/ Bare peat/ Cladonia*
- Bare peat/ *Eriophorum/ Trichophorum*
- *E. vaginatum*

Low ridge (T1):

Oceanic low ridge:
- *Sphagnum/ R. alba*
- *Sphagnum/ Molinia*
- *Schoenus*

Damaged low ridge:
- *E. vaginatum*
- Bare peat/ dwarf shrub
- Bare peat/ *Trichophorum*
- *S. tenellum*

Natural low ridge:
- *S. magellanicum*
- *S. papillosum*

Water's edge (T1/A1):
- *Narthecium/ S. tenellum*
- *S. pulchrum*

* *Calluna, E. nigrum, B. nana, Arctostaphylos, Rubus*

Figure 3(a) Hierarchy of microtope and vegetation stands - Terrestrial. Vegetation stands refer to species constants, though often they are also dominants. In general, these are visually distinct, sometimes striking, stands although obvious cases of co-dominant mixtures also occur (see also Lindsay *et al.* 1985, Lindsay *et al.* 1988). The broad abundance of each stand within each distinct area of pattern type (microtope) should be recorded for comparative evaluation. The list is not comprehensive and other types may be encountered. Work to harmonise the vegetation stands with those used widely in Europe is currently on-going.

Supplement to *Guidelines for selection of biological SSSIs* (Nature Conservancy Council 1989)

Aquatic (A)

Hollows (A1, A2, TA2)

Pools (A3, A4)

Sphagnum hollows (A1)

- S. cuspidatum/ D. anglica
- S. cuspidatum/ R. alba
- S. recurvum
- S. recurvum/ Carex curta
- S. cuspidatum/ Eleocharis multicaulis

Mud-bottom hollows (A2)

- E. angustifolium
- Rhynchorpora alba
- R. fusca
- Carex limosa
- Drosera intermedia/ Molinia

Erosion gullies (TA2)

- Microerosion S. cuspidatum/tenellum/ Trichophorum
- Bare peat/ E. angustifolium

Drought-sensitive pools (A3)

- Menyanthes
- Utricularia/ S. auriculatum
- S. cuspidatum
- S. cuspidatum/ S. auriculatum

Permanent pools (A4)

- Sphagnum/ Menyanthes
- Menyanthes

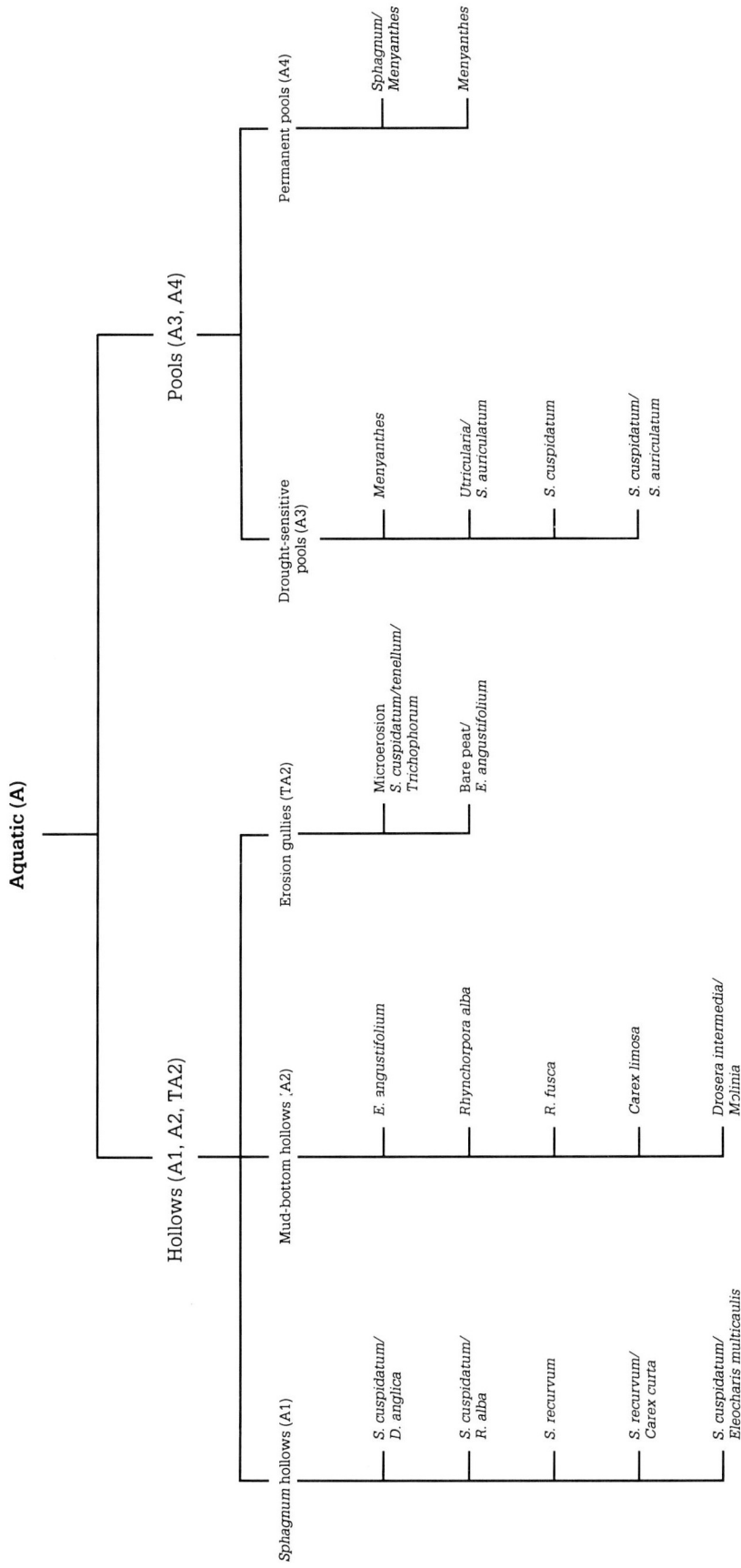

Figure 3(b) Hierarchy of microtope and vegetation stands - Aquatic. Vegetation stands refer to species constants, though often they are also dominants. In general, these are visually distinct, sometimes striking, stands although obvious cases of co-dominant mixtures also occur (see also Lindsay *et al.* 1985, Lindsay *et al.* 1988). The broad abundance of each stand within each distinct area of pattern type (microtope) should be recorded for comparative evaluation. The list is not comprehensive and other types may be encountered. Work to harmonise the vegetation stands with those used widely in Europe is currently on-going.

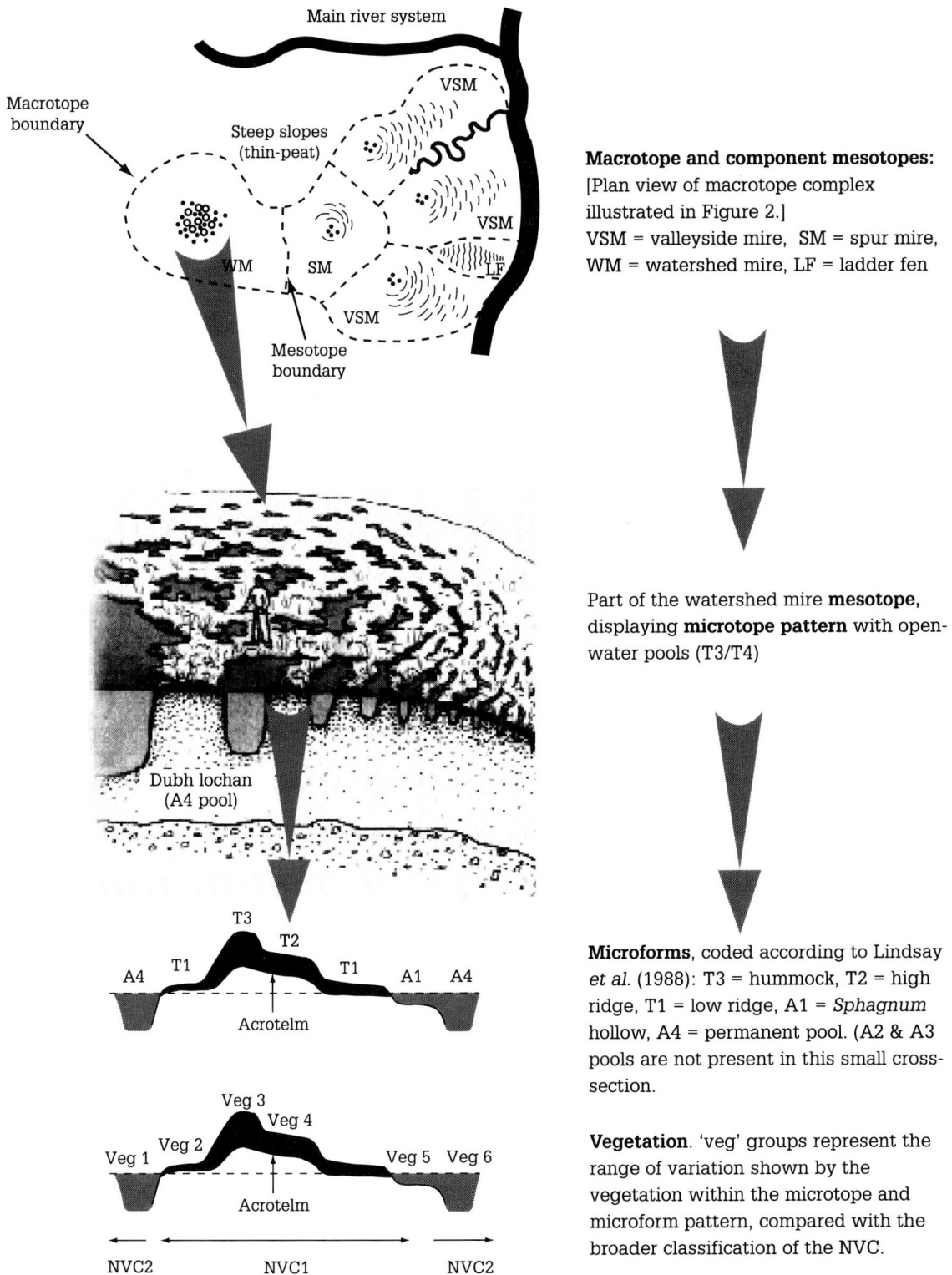

Macrotope and component mesotopes:
[Plan view of macrotope complex illustrated in Figure 2.]
VSM = valleyside mire, SM = spur mire,
WM = watershed mire, LF = ladder fen

Part of the watershed mire **mesotope**, displaying **microtope pattern** with open-water pools (T3/T4)

Microforms, coded according to Lindsay *et al.* (1988): T3 = hummock, T2 = high ridge, T1 = low ridge, A1 = *Sphagnum* hollow, A4 = permanent pool. (A2 & A3 pools are not present in this small cross-section.

Vegetation. 'veg' groups represent the range of variation shown by the vegetation within the microtope and microform pattern, compared with the broader classification of the NVC.

Figure 4   The hierarchy of features used to classify bog systems. Terms are derived from Ivanov (1981) but are described in the accompanying text.

Supplement to *Guidelines for selection of biological SSSIs* (Nature Conservancy Council 1989)

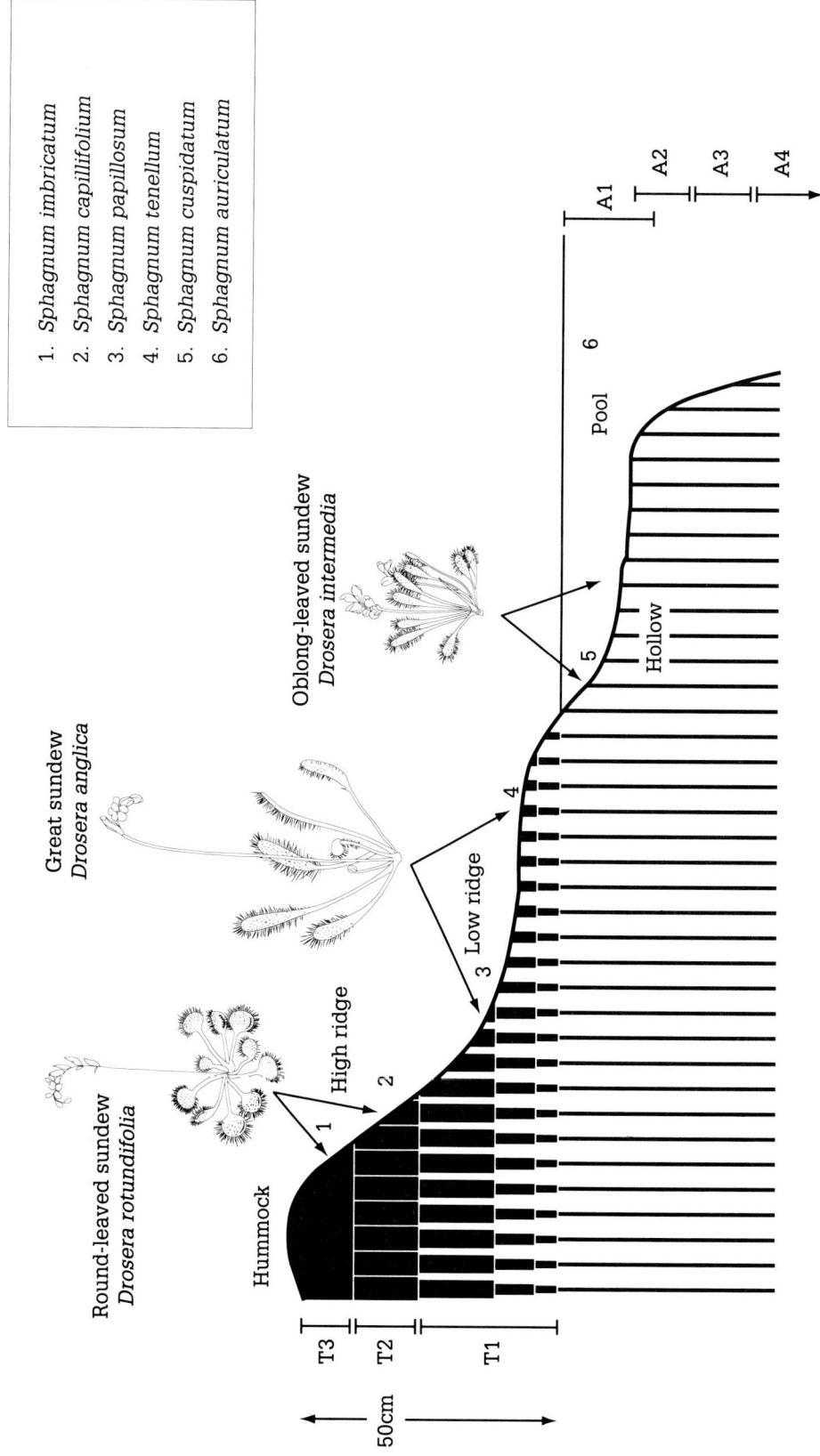

Figure 5   Generalised distribution of structural features (microforms) and the typical distribution of species within the pattern. All natural bogs have some form of pattern, at least across their mire expanse, although in some sites the pattern may consist only of T3 hummocks alternating with T2 high-ridge. Many sites towards the southern and eastern limits of the present bog distribution in Britain have no aquatic (A) zones and consist only of terrestrial (T) zones. (Taken from Lindsay *et al.* 1988)

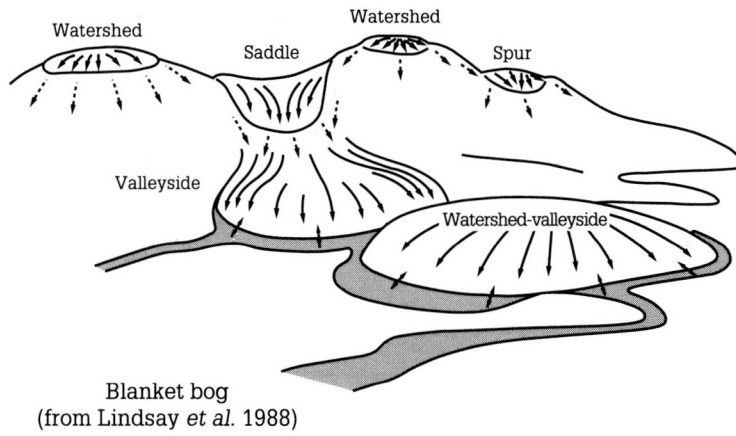

Blanket bog
(from Lindsay *et al.* 1988)

(a)   Blanket bog landscape showing waterflow-lines. Solid flow-lines indicate the flow of water through deep peat; broken lines indicate the flow of water through shallow peat or mineral soil.

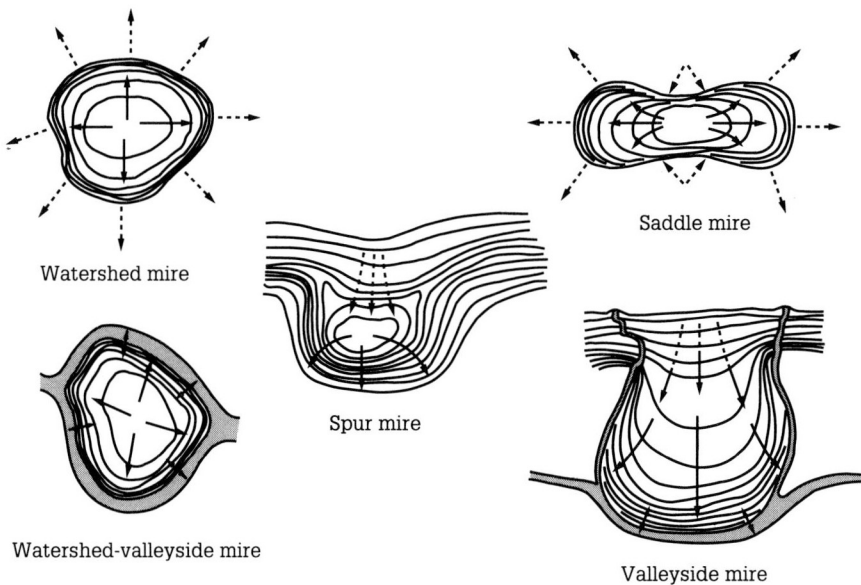

Watershed mire

Saddle mire

Spur mire

Watershed-valleyside mire

Valleyside mire

(b)   Plan views of landscape features showing flow of water through deep peat (solid arrows) and shallow peat or mineral soil (broken lines).

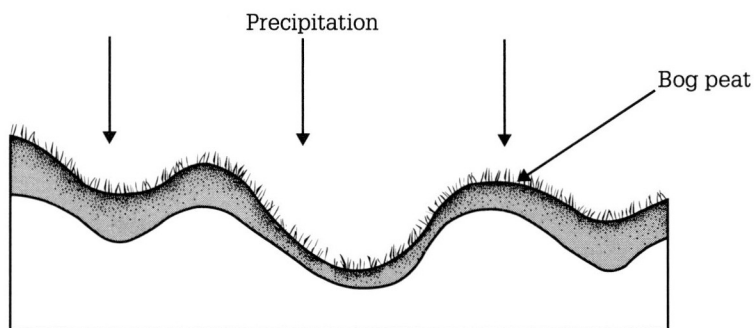

Precipitation

Bog peat

(c)   Sectional view of blanket mire (from Steiner 1992).

Figure 6   The hydromorphological and topographical characteristics of blanket bog

Annex 1. Taken from: *Manual for the interpretation of Annex 1 Priority Habitat Types of the Directive 92/43/EEC* (February 1994, European Commission DG XI B 2).

51.1   * Active raised bogs

habitat code 7110   corine 91 : 51.1
RAISED BOGS AND MIRES AND FENS / SPHAGNUM ACID BOGS

••

1)   **Active raised bogs**

2)   Acid bogs, ombrotrophic, poor in mineral nutrients, sustained mainly by rainwater, with a water level generally higher than the surrounding water table, with perennial vegetation dominated by colourful *Sphagna* hummocks allowing for the growth of the bog (*Erico-Sphagnetalia magellanici, Scheuchzerietalia palustris* p., *Utricularietalia intermedio-minoris* p., *Caricetalia fuscae* p.). Typically, pools may be present in western United Kingdom and Ireland. The term "active" must be taken to mean still supporting a significant area of vegetation that is normally peat forming, but bogs where active formation is temporarily at a standstill, such as after a fire or during a natural climatic cycle eg, a period of drought, are included.

3)   <u>Plants</u>: *Erico-Sphagnetalia magellanici* → *Andromeda polifolia, Carex pauciflora, Cladonia* spp., *Drosera rotundifolia, Eriophorum vaginatum, Odontoschisma sphagni, Shagnum magellanicum, S. imbricatum, S. fuscum, Vaccinium oxycoccos. Scheuchzerietalia palustris* p., *Utricularietalia intermedio-minoris* p., *Caricetalia fuscae* p. → *Carex fusca, C. limosa, Drosera anglica, D. intermedia, Eriophorum gracile, Rhynchospora alba, R. fusca, Scheuchzeria palustris, Utricularia intermedia, U. minor, U. ochroleuca.*
<u>Animals</u>: Dragonflies → *Leucorrhini dubia, Aeshna subartica, A. caerulea, A. juncea, Somatochlora arctica, S. alpestris.* Butterflies → *Colias palaeno, Boloria aquilonaris, Coenonympha tullia, Vacciniina optilete, Hypenodes turfosalis, Eugraphe subrosea.* Spiders → *Pardosa sphagnicola, Glyphesis cottonae.* Ants → *Formica transkaucassia.* Cricket/Grasshopper → *Metrioptera brachyptera, Stethophyma grossum.*

4)   <u>Geographical distribution</u>: Belgium, Denmark, Germany, Spain (Pyrenees and Cantabrian mountains), France, Italy, Ireland, Netherlands and United Kingdom. Variations can occur depending on local climatic and geomorphological conditions. In Belgium, this habitat is only present in High Ardennes; a typical site is the Fagne wallone.
Corresponding category in the United Kingdom National Vegetation Classification: "M1 *Sphagnum auriculatum* bog pool community", "M3 *Eriophorum angustifolium* bog pool community", "M18 *Erica tetralix-Sphagum papillosum* raised and blanket mire", "M20a *Eriophorum vaginatum* blanket and mixed mire - species poor sub community".

5)   In order to support the conservation of this ecosystem over its geographic range and its genetic diversity, marginal areas of lower quality as a result of damage or degradation which abut active raised bogs may need to be included, protected and, where practicable, regenerated. There are very few intact or near-intact raised bogs in Europe

6)   CURTIS, J.R. (in press). The raised bogs of Ireland: their ecology, status and conservation. Government Publications, Dublin.

MOORE, J.J. (1968). A classification of the bogs and wet heaths of northern Europe (Oxycocco-Sphagnetea Br.-Bl. et Tx. 1943). In: Pflanzensoziologische Systematik. Bericht uber das internationale symposium in Stolzenau/Weser 1964 der Internationale vereinigung fur vegetationskunde (R.Tuxen, Ed.). Junk, Den Haag. 306 - 320.

NATURE CONSERVATION COUNCIL (1989). Guidelines for the selection of biological SSSI's. Nature Conservation Council , Peterborough.

SCHOUTEN, M.C.G. (1984). Some aspects of the ecogeographical gradient in Irish ombrotrophic bogs. Peat Congress, Dublin. 1: 414 - 432.

TUXEN, R.; MIYAWAKI, A. & FUJIWARA, K. (1972). Eine erweiterte Gliederung der Oxycocco-Sphagnetea. In: Grundfragen und Methoden in der Pflanzensoziologische. (R.Tuxen, Ed.). Junk, Den Haag. 500 - 520.

52.1 and 52.2   Blanket bog (* active only)

habitat code : 7130   corine 91 : 52.1 and 52.2
RAISED BOGS AND MIRES AND FENS / SPHAGNUM ACID BOGS

....

1)   Blanket bog (* active only)

2)   Extensive bog communities or landscapes on flat or sloping ground with poor surface drainage, in oceanic climates with heavy rainfall, characteristic of western and northern Britain and Ireland. In spite of some lateral water flow, blanket bogs are mostly ombrotrophic. They often cover extensive areas with local topographic features supporting distinct communities [*Erico-Sphagnetalia magellanici*: *Pleurozio purpureae-Ericetum tetracilis*, *Vaccinio-Ericetum tetracilis* p.; *Scheuchzerietalia palustris* p., *Utricularietalia intermedio-minoris* p., *Caricetalia fuscae* p.]. Sphagna play an important role in all of them but the cyperaceous component is greater than in raised bogs.
The term "active" must be taken to mean still supporting a significant area of vegetation that is normally peat forming.

3)   Plants: 52.1 → *Calluna vulgaris*, *Campylopus atrovirens*, *Carex panicea*, *Drosera rotundifolia*, *Erica tetralix*, *Eriophorum vaginatum*, *Molinia caerulea*, *Myrica gale*, *Narthecium ossifragum*, *Pedicularis sylvatica*, *Pinguicula lusitanica*, *Pleurozia purpurea*, *Polygala serpyllifolia*, *Potentilla erecta*, *Racomitrium languginosum*, *Rhynchospora alba*, *Schoenus nigricans*, *Scirpus cespitosus*, *Sphagnum pulchrum*, *S. strictum*, *S. compactum*, *S. auriculatum*. 52.2 → *Calluna vulgaris*, *Diplophyllum albicans*, *Drosera rotundifolia*, *Empetrum nigrum*, *Erica tetralix*, *Eriophorum vaginatum*, *Mylia taylorii*, *Narthecium ossifragum*, *Rubus chamaemorus*, *Scirpus cespitosus*, *Vaccinium myrtillis*.
Animals: *Pluvialis apricaria*, *Calidris alpina*.

4)   Geographical distribution: France, Ireland and United Kingdom.
Sub-types of the British Isles: 52.1 → Hyper-Atlantic blanket bogs of the western coastlands of Ireland, western Scotland and its islands, Cumbria, Northern Wales; bogs locally dominated by sphagna (*Sphagnum auriculatum*, *S. magellanicum*, *S. compactum*, *S. papillosum*, *S. nemoreum*, *S. rubellum*, *S. tenellum*, *S. subnitens*), or, particularly in parts of western Ireland, mucilaginous algal deposits (*Zygogonium*). 52.2 → Blanket bogs of high ground, hills and mountains in Scotland, Ireland, Western England and Wales.
Corresponding category in the United Kingdom National Vegetation Classification: "M1 *Sphagnum auriculatum* bog pool community", "M15 *Scirpus cespitosus-Erica tetralix* wet heath", "M17 *Scirpus cespitosus-Eriophorum vaginatum* blanket mire", "M18 *Erica tetralix-Sphagnum papillosum* raised and blanket mire", "M19 *Calluna vulgaris-Eriophorum vaginatum* blanket mire", "M20 *Eriophorum vaginatum* blanket mire".

5)   In the United Kingdom discrete areas of raised bog and blanket bog may occur in some districts, showing their characteristic differences. In many other areas, however, peatlands which may have begun as raised bog have became merged in a general expanse of blanket bog, losing their distinctive marginal features. Within these blanket bogs, there are other peat-forming systems which, strictly speaking, form part of various biotopes of aquatic and amphibious zones, fens and moorland.

6)   DOYLE, G.J. & MOORE,J.J. (1980). Western blanket bog (Pleurozio purpureae-Ericetum tetralicis) in Ireland and Great Britain. Colloques Phytosociologiques. VII: 213 - 223.

MOORE, J.J. (1968). A classification of the bogs and wet heaths of northern Europe (Oxycocco-Sphagnetea Br.-Bl. et Tx. 1943). In: Pflanzensoziologische Systematik. Bericht uber das internationale symposium in Stolzenau/Weser 1964 der Internationale vereinigung fur vegetationskunde (R.Tuxen, Ed.). Junk, Den Haag. 306 - 320.

NATURE CONSERVATION COUNCIL (1989). Guidelines for the selection of biological SSSI's. Nature Conservation Council , Peterborough.

TUXEN, R.; MIYAWAKI, A. & FUJIWARA, K. (1972).   Eine erweiterte Gliederung der Oxycocco-Sphagnetea. In: Grundfragen und Methoden in der Pflanzensoziologische. (R.Tuxen, Ed.). Junk, Den Haag. 500 - 520.